# Emerging Technologies in Clean Energy

Topical Conference at the 2023 AIChE Spring Meeting and 19th Global Congress on Process Safety

Houston, Texas, USA
12-16 March 2023

ISBN: 978-1-7138-7934-3

**Printed from e-media with permission by:**

Curran Associates, Inc.
57 Morehouse Lane
Red Hook, NY 12571

**Some format issues inherent in the e-media version may also appear in this print version.**



Printed with permission by Curran Associates, Inc. (2023)

For permission requests, please contact AIChE
at the address below.

AIChE
120 Wall Street, FL 23
New York, NY 10005-4020

Phone: (800) 242-4363
Fax: (203) 775-5177

www.aiche.org

**Additional copies of this publication are available from:**

Curran Associates, Inc.
57 Morehouse Lane
Red Hook, NY 12571 USA
Phone: 845-758-0400
Fax: 845-758-2633
Email: curran@proceedings.com
Web: www.proceedings.com

# TABLE OF CONTENTS

## CO2 CAPTURE, UTILIZATION, AND DISPOSAL: KEY TO CLEAN ENERGY PRODUCTION II



## ALTERNATIVE FUELS INCLUDING BIOFUELS, HYDROGEN, RENEWABLE HYDROGEN, AND SYNGAS II

# 92a Development of a Rigorous Electrolyzer Model for Performing Robust, Dynamic Studies

*Brent Bishop*

In the 1990s, a push for cleaner and more efficient power-generating processes led to the development of technologies such as Integrated Gasification Combined Cycle (IGCC). They were developed to produce cleaner and more efficient electricity as well as hydrogen fuel from the gasification of coal or biomass[1]. Throughout the 2000s, IGCC gradually improved and showed potential to compete both economically and environmentally with the alternative Natural Gas Combined Cycle (NGCC) [2]. This trend changed in 2011 when discovery of a large natural gas source in the Marcellus Shale[3] led to increased fracking and thus reduced natural gas prices, leaving hydrogen processes based on the existing coal supply (like IGCC) unable to compete. However, societal pressure centered around addressing rising global temperatures has created a renewed interest in pursuing the cleaner, hydrogen-fueled economy of the 2000s.

Although steam-methane reforming is the main producer of hydrogen (referred to as "blue" hydrogen), another route being investigated is hydrogen produced through the electrolysis of water[4]. This process uses electricity from the grid to split water into hydrogen fuel and oxygen. The benefit of this approach is that the electricity can come from completely renewable sources and involves no carbon (unlike steam-methane reforming), making this a "green" hydrogen process. It is also a promising technology as it allows for the chemical storage of renewable energy in the form of fuel, an alternative to battery storage. Because these electrolyzers are connected to a renewable grid that is inherently dynamic, there is a need to develop detailed, robust, and easy-to-use dynamic models of electrolyzers.

In this work, a rigorous electrolyzer model is proposed and developed in AVEVA™ Process Simulation's equation-oriented environment[5]. The electrodes for the cell are modeled as two tanks and allow for the calculation of material as well as energy holdup inside the electrolyzer. This is crucial for accurately simulating safety scenarios where the source of water is cutoff, and the water level begins to lower. Accurate energy calculations are also important as many of the material properties that determine voltage like electrical resistance are temperature-dependent[6], [7]. These tanks are separated by a semi-permeable membrane where $H^+$ and $OH^-$ ions can be transferred from anode to cathode and vice versa. Lastly, a user-defined, replaceable submodel can be defined to describe the voltage of a given electrolyzer. This gives the user the ability to customize the electrolyzer model with as much detail as they require for a given study.

With this model, P&ID-level detail can be achieved for electrolyzer-focused dynamic simulations. By simply defining a voltage submodel and the design parameters of their electrolyzer, a user can quickly and seamlessly begin testing control systems, perform technoeconomic analyses, and conduct safety studies. The ability to perform such analyses is necessary for transitioning to cleaner alternative fuels like green hydrogen over the coming years.

**References**

[1] R. G. Lundqvist, "The IGCC demonstration plant at Värnamo," *Bioresour. Technol.*, vol. 46, no. 1–2, pp. 49–53, Jan. 1993, doi: 10.1016/0960-8524(93)90053-E.

[2] G. Ordorica-Garcia, P. Douglas, E. Croiset, and L. Zheng, "Technoeconomic evaluation of IGCC power plants for CO2 avoidance," *Energy Convers. Manag.*, vol. 47, no. 15–16, pp. 2250–2259, Sep. 2006, doi: 10.1016/j.enconman.2005.11.020.

[3] "Assessment of Undiscovered Oil and Gas Resources of the Devonian Marcellus Shale of the Appalachian Basin Province, 2011." https://pubs.usgs.gov/fs/2011/3092/ (accessed Nov. 10, 2022).

[4] E. R. Morgan, J. F. Manwell, and J. G. McGowan, "Opportunities for economies of scale with alkaline electrolyzers," *Int. J. Hydrog. Energy*, vol. 38, no. 36, pp. 15903–15909, Dec. 2013, doi: 10.1016/j.ijhydene.2013.08.116.

[5] AVEVA Group Plc., "Product Datasheet: AVEVA Process Simulation." AVEVA, 2020. Accessed: Nov. 10, 2022. [Online]. Available:
https://www.aveva.com/content/dam/aveva/documents/datasheets/Datasheet_ProcessSimulation_EN.pdf.coredownload.inline.pdf

[6] O. Ulleberg, "Modeling of advanced alkaline electrolyzers: a system simulation approach," *Int. J. Hydrog. Energy*, vol. 28, no. 1, pp. 21–33, Jan. 2003, doi: 10.1016/S0360-3199(02)00033-2.

[7] Z. Abdin, C. J. Webb, and E. MacA. Gray, "Modelling and simulation of an alkaline electrolyser cell," *Energy*, vol. 138, pp. 316–331, Nov. 2017, doi: 10.1016/j.energy.2017.07.053.

## 92b Anaerobic Biodegradability and Potential Recovery of Biomethane from Agro-Agricultural Wastes

*Harjinder Kaur, Raghava Kommalapati, Faryal Fatima*

The anthropogenic activities especially fossil fuel-driven climate change fuel the sustainable initiatives for waste management and renewable energy. The anaerobic co-digestion of goat manure with sorghum, cotton gin trash, and food waste having different mixing ratios, volumes, temperatures, and additives was optimized in single and two-stage bioreactors. The biochemical methane potential assays (having different mixing ratios of double and triple substrates) were run in triplicates in 250 mL serum bottles. The best yielding ratios were up-scaled in fabricated 2-L bioreactors. The biodegradabilities, biomethane recoveries, and process efficacies are discussed. The co-digestion of goat manure with sorghum in a 70:30 ratio yielded the highest biomethane, which was further up-scaled to a two-stage temperature-phased process having varied nutritional conditions in fabricated 2-L bioreactors. This system yielded the highest biomethane of 266.0 $mL/g_{VS}$ having an anaerobic biodegradability of 67.3%. The resulting effluent fractions; liquid and solids were separated. The liquid fraction was standardized for struvite precipitation and analyzed for plant available nitrogen and phosphate. The solid fraction was analyzed for several macro and micronutrients for application as biofertilizers. This study involving feedstock acquisition, characterization and their anaerobic digestion optimization provides fundamental information and may assist small farmers operating on-farm anaerobic digesters.

# 92c Hydrogen FUEL – a Friend to the Environment or a Foe to Human Lives?

*Kumar Israni*

In today's times, all industries are headed towards alternate energy sources, while contributing to their portion of caring for world climate changes in environment. The emphasis on Low Carbon emissions has been tremendous. One of these options is the emerging use of hydrogen as a source of fuel. The hydrogen molecule is nothing new to the process industry. It's properties and applications have been in existence since its discovery back in 1766.

As environmental specialists would agree, while reviewing the energy value chain, hydrogen has tremendous advantages over other alternatives, by being a continuously available resource. Use of hydrogen as fuel generates ultra-low emissions and hence has earned the status of clean energy.

As process safety professionals, our focus is always around safe operations, and the emphasis on the value of human lives, so that all of us make it back to our loved ones safe and sound, every day. In light of this mission, the chemical element "hydrogen" does pose tremendous threats to human lives while its application as a fuel. There have been incidents ascertaining this threat, and also lessons learned.

This paper will discuss some current findings from incidents, and the author's views on what should be this balance between environment and safety. This will incorporate the current regulatory stance on hydrogen safety, and how will that impact the balance between the environment and safety; while continuing the worldwide mission of low carbon, that in turn benefits climate change.

## 144a Chemical Capture of CO2 and Conversion to Alkyl Carbonates from the Smr Based Production of Blue Hydrogen

*Chandrakant (CB) Panchal, Christopher L. Marshall, Richard Doctor, Dennis Miller*

Hydrogen has become a crucial link to the 2050 carbon-neutrality goal. Industry is actively pursuing Steam Methane Reforming (SMR) of natural gas (NG) and renewable natural gas (RNG). The industry is pursuing energy efficient SMR process by heat integration for hydrogen production with low carbon-footprint compared to commercial SMR. $CO_2$ emissions from conventional SMR process is more than 7 kg $CO_2$ / kg $H_2$; therefore, achieving the goal of 1 kg $CO_2$/kg $H_2$ requires $CO_2$recovery of 90%. Furthermore, to enhance the economics of $H_2$ and achieve target cost of $H_2$, the recovered $CO_2$ must be converted to high-value products with expanding global market demands. Process analysis is performed of chemical $CO_2$ capture and conversion to alkyl carbonates for achieving both of these goals. A trade-off analysis is performed for chemical capture of $CO_2$ from three locations in the SMR process for 90% capture with favorable economics. The process development builds upon E3Tec's process development of an integrated process of $CO_2$ capture and its conversion to high-purity alkyl carbonates by alternate chemical pathways. E3Tec has developed a conceptual adsorbent catalytic reactor (ACR) using an integrated solid adsorbent and catalyst for direct capture from the primary $CO_2$ source and its conversion to high value alkyl carbonates. Case studies consist of conversion of ethylene or propylene oxide to their corresponding carbonates with further conversion dimethyl carbonate (DMC) with selective coproduction of glycols using Heat-Integrated Reactive Distillation (HIRD) process. The formation of high-value alkyl carbonates, with an increased need in the expanding market of lithium-ion batteries, will have positive impact on $H_2$ economics with low carbon-footprint.

## 144b An Experimental Study of a Microwave-Assisted Direct Air Capture.

*Mustafa Erguvan, Matthew Atwood, Shahriar Amini*

**An Experimental Study of a Microwave-Assisted Direct Air Capture.**

Mustafa Erguvan[1], Matt Atwood[2], Shahriar Amini[1]

1. University of Alabama, Department of Mechanical Engineering, Tuscaloosa, AL, USA
2. AirCapture, LLC, CA, USA

# Abstract

Over 190 countries have agreed to reduce greenhouse gas emissions as well as limit the global temperature increase to 2 °C through the Kyoto Protocol and the Paris Agreement. To fulfill the requirements of this agreement, negative emission technologies will play an important role in the next decades. Direct Air capture (DAC) systems, which is defined simply as removal of $CO_2$ from the atmosphere, has been considered as one of the most promising ways to reduce and maintain the $CO_2$ concentration of the air. Although several experimental and numerical studies related to the DAC system have been performed, the main feasibility challenge for implementation of DAC systems is the high capital and energy costs of $CO_2$ capture.

There are two main DAC systems namely, high temperature aqueous solution and low temperature solid sorbent. While high temperature systems require over 800 °C for the desorption process, 80-100 °C could be sufficient using solid sorbent technology, which makes the low temperature system more advantageous since even waste heat from industrial plants can be used as a heat source.

As the reaction binding $CO_2$ to sorbents is exothermal in adsorption, the addition of thermal energy is required to regenerate the sorbent and capture the $CO_2$ during desorption. Conventionally used desorption systems include temperature swing adsorption (TSA), pressure swing adsorption (PSA) and vacuum swing adsorption (VSA) where desorption of $CO_2$ is normally done by changing the temperature and pressure (TVSA). This desorption step represents the majority of overall DAC system capital cost and energy demand, although an argument is often made that leveraging waste heat reduces desorption costs; however, this approach limits the market potential of DAC by requiring co-location with waste thermal heat sources and requires costly capital equipment associated with steam production, management and thermal recovery, while impacting overall system carbon lifecycle [1]. Therefore, approaches that can target desorption energy to the sorbent-$CO_2$ bond while limiting or avoiding the sensible heat loss associated with the contactor substrate and related DAC component parts can maximize second law efficiency leading to lower.

The use of microwave (MW) irradiation, replacing the conventional heating technique, can result in process intensification, faster scale up and energy costs reduction. The electromagnetic energy is directly converted into thermal energy inside the adsorbent, overcoming the heat transfer limitations. Several researchers have explored the use of MW for $CO_2$ desorption[2-7].

In this study, a microwave assisted regeneration technique has been used as a heat source to capture

$CO_2$ with a solid contactor which was made of a titanium oxide monolith impregnated with polyethylenimine (PEI). A modified microwave oven was used, heating the contactor to temperatures below 130 °C.

Several parametric studies have been performed to investigate the adsorption and desorption processes by evaluating $CO_2$ concentration and temperature of the released gas. While microwave power varied between 336 W and 840 W, the volumetric flow rate of the inlet gas changed from 40 ml/min to 100 ml/min. Furthermore, repeated experiment cycles with the same conditions have also been performed to evaluate the adsorption and desorption stabilities of the contactor. The results showed that the microwave power has more effect than inlet gas flow rate on the desorption process.

The use of microwave has shown to be effective in $CO_2$ desorption of amino-polymer sorbents on the contactor. This addresses the heat transfer limitations occurring in conventional thermal desorption processes.

# Acknowledgment

The funding from Alabama Transportation Institute (ATI) is acknowledged to carry out this work.

# References

1. Fasihi, M., O. Efimova, and C. Breyer, *Techno-economic assessment of CO2 direct air capture plants.* Journal of Cleaner Production, 2019. **224**: p. 957-980.
2. Chronopoulos, T., et al., *Utilisation Of Microwave Energy for CO2 Desorption in Post-combustion Carbon Capture Using Solid Sorbents.* Energy Procedia, 2014. **63**: p. 2109-2115.
3. McGurk, S.J., et al., *Microwave swing regeneration of aqueous monoethanolamine for post-combustion CO2 capture.* Applied Energy, 2017. **192**: p. 126-133.
4. Bougie, F. and X. Fan, *Microwave regeneration of monoethanolamine aqueous solutions used for CO2 capture.* International Journal of Greenhouse Gas Control, 2018. **79**: p. 165-172.
5. Tsubaki, S., et al., *Insights into the Dielectric-Heating-Enhanced Regeneration of CO2-Rich Aqueous Amine Solutions.* ACS Sustainable Chemistry and Engineering, 2020. **8**(36): p. 13593-13599.
6. Meloni, E., et al., *Intensification of TSA processes using a microwave-assisted regeneration step.* Chemical Engineering and Processing - Process Intensification, 2021. **160**.
7. van Schagen, T.N., P.J. van der Wal, and D.W.F. Brilman, *Development of a novel, through-flow microwave-based regenerator for sorbent-based direct air capture.* Chemical Engineering Journal Advances, 2022. **9**: p. 100187.

## 144c Renewable Energy to Synthetic Natural Gas Pipeline

*Yoshitsugi Kikkawa, Alejandra Peralta, David Hill, Noelle Garza*

Due to increasing concerns about global warming, the industry has constantly been looking for alternative technologies to minimize carbon emissions to the environment. The utilization of renewable energy, such as wind turbine power, solar energy, and others, has become essential and popular nowadays.

In previous studies, we proposed an approach to take advantage of green hydrogen and captured CO2 to produce Synthetic Natural Gas (SNG) by the Sabatier reaction. When the SNG is not used nearby the production facility, transportation becomes another potential source of carbon emissions.

This paper studies the pipeline design to transport Synthetic Natural Gas (SNG) inland in places like North America, South America, and Eurasia. We modeled a 1,000 km to 10,000 km SNG pipeline using CHEMCAD simulation software. The result of this study shows close to zero emissions to the atmosphere.

## 45a Rare Earth Metals Recycling from Electric Vehicles

*Maoqi Feng*

Electric vehicles will be dominating in future transportation market. Rare earth metals (REMs) are crucial to the production of green energy motors in electric vehicles. Recycling of valued-added metals especially strategic metals such as REMs is necessary for sustainable global economy in the near future. This paper will present our recent studies on REMs recycling from electromagnetic motors using REMs as permanent magnets. A life cycle analysis and a techno-economic analysis will also be presented.

# 45b Ocean Thermal Energy in the Gulf of Mexico for Coproduction of Desalinated Water and Green Ammonia As Hydrogen Carrier

*Chandrakant (CB) Panchal, Luis Vega*

The world's oceans are the largest collectors and storage of the solar energy. Large scale deployment of Ocean Thermal Energy Conversion (OTEC) plants in tropical and subtropical regions will be a Bridge to 2050 Decarbonization goal. Deployment of large scale OTEC plantships can have positive impacts on ocean environment beside producing desalinated water and green ammonia as hydrogen carrier. This is specifically true for the Gulf of Mexico, which has been one of the favourable sites for OTEC plants. The critical aspect of large-scale OTEC plants in the Gulf of Mexico is to extract the energy from the surface seawater and discharge mixed seawater at an optimal depth. The extraction of thermal energy from the surface water and upwelling of deep-ocean cold water would reduce systematic rise of the surface seawater temperature of the Gulf of Mexico. This is a visionary and ambitious goal. However, in the recent years, surface temperature of the Gulf of Mexico has increased by 1-2°C, causing severe storms with unpredictable intensity, in addition to acidification of seawater. Large-scale OTEC plants will moderate rising of Gulf of Mexico seawater temperature, beside direct impact on the abatement of green-house gas emissions from the Gulf States. The Gulf of Mexico is a major producer of oil and natural gas with more than 1,700 oil platforms and rigs. At current production rate of about 1.6 million barrels per day, the oil reserve will start depleting in coming decades that will cause decommissioning of offshore oil platforms. The present assessment study focuses on the following key elements of large-scale deployment of OTEC plants in the Gulf of Mexico: a) production of ammonia as hydrogen carrier; b) freshwater production for the water-stressed Caribbean Island and Gulf States; c) conversion of captured carbon dioxide from the Gulf States to specialty and commodity chemicals, such as methanol, dimethyl ester and alkyl carbonates; and d) recovery and conversion of carbon dioxide from seawater. An analysis was performed for production of 3,000 tonnes per day of ammonia along with desalination of seawater to evaluate the techno-economic assessment (TEA) of satellite OTEC plantships by conversion of oil platforms to ammonia producing plants. The electric power to the ammonia synthesis process is supplied by floating OTEC plants. Emerging energy efficient ammonia synthesis processes enhances the overall techno-economic merits. Conversion of out-of-service oil platforms to OTEC plant significantly reduces the total-installed-cost (TIC) by taking the credit for costs associated with dismantling and disposal of oil platforms. The Technology Readiness Level (TRL) is at such a stage that such a satellite OTEC plantships can be deployed in foreseeable future with low techno-economic risks.

# 45c Waste Motor Oil Pyrolysis for Clean Energy Generation in a Self Designed Pyrolyzer

*Asmita Mishra, Bhim Charan Meikap*

## Background and Motivation

Motor oil is an engine lubricant for heat transfer and friction reduction. But with use due to alteration in viscometric properties, they become unsuitable for further use and are discarded as waste. This discarded waste oil is detrimental to human health and the environment due to its carcinogenic, mutagenic, and regenerative properties. Thus, it demands sustainable waste management technology. Due to its high heating value, this waste oil can be used for clean energy generation using eco-friendly pyrolysis technology. Pyrolysis has gained attention for its flexibility and diverse product ranges.

## Scope of the Study

This study assesses the feasibility of using WMO for clean energy generation in a self-designed pyrolyzer. Besides, the generated pyro-oil characteristics are compared with the diesel fuel properties.

## Materials and Methods

The WMO was procured from the Tech Market within the Indian Institute of Technology Kharagpur (22.3 °N 87.3 °E), India premises. The feedstock was stored in an air-tight container for further use.

The designed pyrolyzer can be altered according to the researcher's demand. However, in the present study, only the semi-batch approach is explored to optimize the pyro-oil yield. The overall efficiency of the pyrolysis process is influenced by its process parameters. The process parameters considered in this study are reactor temperature ($T_R$, °C), heating rate (β, °C/min), nitrogen flowrate (v, LPM), and residence time (t, min). Besides, COMSOL Multiphysics was utilized to understand the rate of pyrolysis within the reactor. Lastly, the produced pyro-oil characteristics are compared with diesel fuel.

## Results

The results show a continuous increase in pyro-oil yield from 9 to 82 wt% with an increase in $T_R$ from 300 to 500 °C and a decrease in residue from 86.1 to 11 wt%. Similarly, an increase in pyro-oil yield from 82 to 90 wt% and a decrease in residue from 10.8 to 2.4 wt% was observed with an increase in heating rate from 4 to 8 °C/min. However, with an increase in v and t, there is an increase in operational costs. So, balancing the economics with product yield, $T_R$ = 500 °C, β = 8 °C/min, n = 0.6 LPM and t = 10 min is selected as the optimum process parameters. This increase in the decomposition is mainly due to the secondary reactions leading to a greater yield of the pyrolysis products. The COMSOL modelling, helps understand the reactor's temperature distribution and its co-relation with the product yield. Besides, the fuel properties of the generated pyro-oil are in line with commercial diesel fuels.

## Implications

The finding of this research article will help design and develop a continuous and cost-effective pyrolyzer for recovering alternate energy, thereby increasing the sustainability of WMO pyrolysis in the future. Besides, the generated pyro-oil can be used as a transport fuel because of its diesel-like fuel properties.

## 45d Integrated Planning of Low-Iluc Biomass Production and Supply Chains for Advanced Biofuels: Towards the EU Biobased Economy

*Dauda Ibrahim, Sara Giarola, Calliope Panoutsou, Nilay Shah*

The increased concern over climate change and the gradual depletion of fossil fuel have led to the search for alternative sources of energy that are both renewable and sustainable. Biofuels and bioliquids from biomass have been used in the transport sector (for aviation, road transport, and maritime) as alternatives to fossils to decarbonise the sector. One of the major drawbacks is the increased demand for land use to cultivate biorefinery feedstock such as biomass for advance biofuel production. To avoid competition with land used for food production, the EU have put in place policies that mandate sustainable biomass supply through introduction of low-indirect land use change (low-ILUC). Low-ILUC risk biofuel are produced from feedstock that avoid food and feed crop displacement through (i) yield increases from improved agricultural practices or (ii) cultivation on areas not previously used for crop production, for example, unused, abandoned or severely degraded land. To speed-up market uptake and certification by international sustainability and carbon certification (ISCC), there is a need for a systematic approach that can assess the benefit of both existing and emerging low-ILUC risk feedstock, for example *Brassica Carinata, Perennial crops, Castor oil,* etc.

This work proposes a novel and systematic approach to plan the production of low-ILUC risk feedstock, with the aim of designing an advanced biofuel supply chain including the most profitable crop management practice whilst aiming to a minimum ILUC. The approach integrates an optimisation model for crop production that incorporates crop rotation and a mixed-integer linear programming model for the entire biofuel value chain, which includes sourcing of low-ILUC biomass feedstock, biofuel production, and distribution of biofuel from plants to market. The inputs to the crop production model use real-world data for a set of crops, crop management options, and available land, which include fertilizer and energy consumption, market price, production cost, crop yield, and water needs. The model aims at optimising the annual land allocated to each crop/crop sequence over the entire planning horizon considering minimising the risk for ILUC. Likewise, the biofuel value chain model consists of inputs required to estimate capital cost, operating cost, transportation cost, and revenue. The overall objective is to maximise farmers' profit while satisfying biofuel demand in both local and international markets. Farmers' profit is maximised with the inclusion of carbon credits derived from optimal carbon practices and by maximising land productivity, which corresponds to the minimum ILUC. The economic analysis incorporate cost related to cultivation of biomass, production of biofuel, and distribution of biofuels.

The capability of the proposed systematic approach is demonstrated using an industrial case study with Brassica napus. Outcomes from this study define benchmarks of crop rotation and crop additionality in the European regulatory context of low-ILUC risk status for use in biofuel from 2020 to 2030; inform primarily the bioenergy and biofuels but also other bioproduct sectors (biochemicals/biopolymers, chemical industry); support the sustainable conversion of the chemical industry; provide policy and market stakeholders with new knowledge; and to remove the most prominent barriers against the market uptake of low-ILUC risk biofuels, bioliquids and biomass fuels.

## 45e Improving Biochar Surface Area for $CO_2$ Adsorption through Potassium Hydroxide Chemical Activation

*Conner Pope, Tawsif Rahman, Gabriel A. Goenaga, Sushil Adhikari, Thomas Zawodzinski, Nicole Labbé, Nourredine Abdoulmoumine*

With climate change continuing to worsen, developing low-cost, sustainable adsorbents is critical for implementing carbon dioxide ($CO_2$) capture on an industrial scale. Biochar is a solid material formed from the thermal decomposition of lignocellulosic biomass in an oxygen-deprived environment. This material can be applied as a low cost, sustainable adsorbent for capturing $CO_2$ but requires surface area improvement before being used. In this work, loblolly pine biochar created from intermediate pyrolysis was chemically activated with potassium hydroxide (KOH) at different activation temperatures, holding times, and KOH:biochar mass ratios to maximize biochar surface area for $CO_2$ adsorption. Thermogravimetric analysis (TGA) demonstrated that activation improved biochar thermal stability and removed degraded volatile compounds retained in biochar following pyrolysis. By nitrogen ($N_2$) physisorption, this evolution of volatiles increased biochar surface area for most activation condition combinations with biochar activated at 800°C for 3 h with a 1:1 KOH:biochar mass ratio having the highest specific surface area of 1080 $m^2/g$. X-ray diffraction (XRD) revealed that biochar aromatic sheet size increased with increasing activation temperature from 16 in biochar to 21-36 Å in activated biochar. Fourier-transform infrared (FTIR) spectroscopy showed that aliphatic and oxygenated biochar functionalities were removed with increasing activation temperature. For $CO_2$ adsorption, the highest surface area activated biochar reached breakthrough and regeneration later than biochar reflecting its significantly higher surface area. As such, $CO_2$ adsorption capacity increased from 0.45 to 0.73 mmol/g through activation representing a 62% improvement in carbon capture.

SPRING23
+19TH GCPS
A Joint AIChE and CCPS Meeting

# Application of Combustion Synthesized $La_{0.5}A_{0.5}MnO_3$ (where, A = Ca, Mg, Sr, Ba) for Thermochemical Production of Fuel from $CO_2$

**(Principal Author)**
**Rahul Bhosale**
Department of Chemical Engineering,
University of Tennessee, Chattanooga, TN, 37421, USA
rahul-bhosale@utc.edu
rrbhosle1985@gmail.com

**(Co-Authors)**
**Sayma Akhter**
Department of Chemical Engineering,
Qatar University, Doha, Qatar

**Vipul Patil**
Department of Mechanical Engineering,
University of Tennessee, Chattanooga, TN, 37421, USA

**Gabrielle Bennett**
Department of Chemical Engineering,
University of Tennessee, Chattanooga, TN, 37421, USA

Prepared for Presentation at
American Institute of Chemical Engineers
2023 Spring Meeting and 19th Global Congress on Process Safety
Houston, TX, USA
March 12 – 16, 2023

# Application of Combustion Synthesized $La_{0.5}A_{0.5}MnO_3$ (where, A = Ca, Mg, Sr, Ba) for Thermochemical Production of Fuel from $CO_2$

**Rahul R. Bhosale*[,1], Sayma Akther[2], Vipul Patil[3], Gabrielle Bennett[1]**

[1]Department of Chemical Engineering, University of Tennessee, Chattanooga, TN, 37421, USA
[2]Department of Chemical Engineering, Qatar University, Doha, Qatar
[3]Department of Mechanical Engineering, University of Tennessee, Chattanooga, TN, 37421, USA

**Keywords:** La-based Perovskites, Thermochemical, $CO_2$ Utilization

**Abstract**

This work attempts to dope multiple alkaline earth metal cations (Ca, Mg, Sr, and Ba) in the lanthanum-manganese perovskite crystal structure. A solution combustion synthesis approach is used to develop $La_{0.5}A_{0.5}MnO_3$ (where A = Ca, Mg, Sr, Ba). The pre-reaction characterization of the derived $La_{0.5}A_{0.5}MnO_3$ perovskites is carried out using powder x-ray diffraction and scanning electron microscope. Furthermore, a high-temperature thermogravimetric analyzer-based reactor setup is developed to test the redox ability of the derived $La_{0.5}A_{0.5}MnO_3$ perovskites.

## 1. Introduction

Fig. 1 shows the annual $CO_2$ emissions of the world. The data presented shows a significant concern regarding the release of $CO_2$ as it is responsible for the greenhouse effect and, ultimately, global warming. Much research has been done on the capture of $CO_2$ from industrial exhaust sources [1,2]. On the other hand, in recent years, much focus has shifted toward using $CO_2$. One promising research area focusing on converting $CO_2$ into fuels is the solar thermochemical cycles driven by metal oxide-based redox reactions. Volatile such as zinc oxide[3–9] and tin oxide[10–15], and non-volatile such as ferrites[16–31], doped ceria[32–45], and perovskites[46–54] are the two types of metal oxides are applied for the solar thermochemical cycles in the past few years.

The solar thermochemical community has recently shifted towards utilizing La-based perovskites. Recently, our team published an article in which we synthesized La-Sr-Mn-O perovskite using the combustion synthesis method and tested these perovskites for the $CO_2$ splitting application. The results obtained are promising; however, there is a scope for improvement in fuel production and cyclability. Hence, in this study, we further doped the La-Mn-O perovskite by alkaline earth metal cations (Ca, Mg, Sr, Ba) to compare their applicability towards converting $CO_2$ into fuels. The redox reactions associated with the $La_{0.5}A_{0.5}MnO_3$ perovskite-based conversion of $CO_2$ are explained as follows:

$$La_{0.5}A_{0.5}MnO_3 \rightarrow La_{0.5}A_{0.5}MnO_{3-\delta_{red}} + \frac{\delta_{red}}{2}O_2 \qquad (1)$$

$$La_{0.5}A_{0.5}MnO_{3-\delta_{red}} + (\delta_{red})CO_2 \rightarrow La_{0.5}A_{0.5}MnO_3 + (\delta_{red})CO \qquad (2)$$

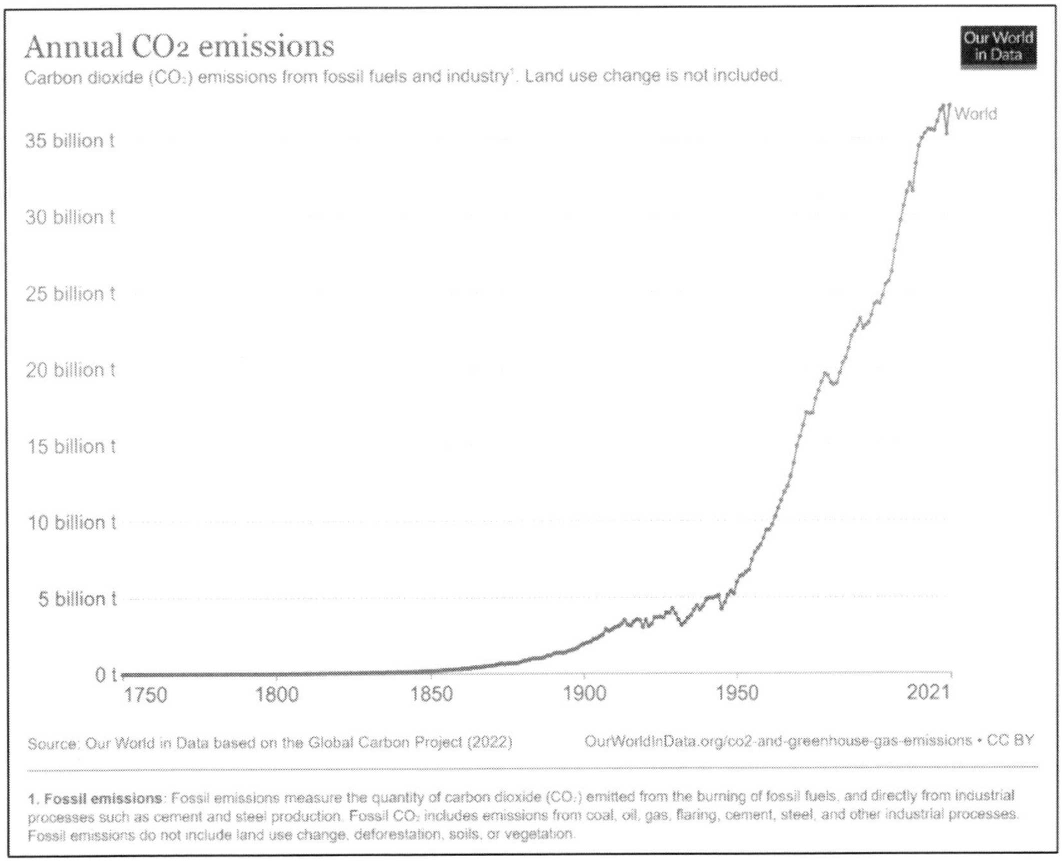


Fig. 1. Annual $CO_2$ emissions associated with fossil fuel utilization.

## 2. Experimental

The solution combustion synthesis requires the following precursors for the synthesis of $La_{0.5}A_{0.5}MnO_3$ perovskites: nitrates of La, Mn, Ca, Mg, Sr, and Ba (purchased from Sigma Aldrich, USA), glycine as a fuel, and deionized water. Following glassware and supplies are needed: beakers, hot plate, thermocouple, and mesh. The first step associated with the

The first step associated with synthesizing $La_{0.5}A_{0.5}MnO_3$ perovskites via the combustion method is dissolving the metal nitrate in water with the help of a magnetic stirrer. The weights of the precursors are calculated by considering the reaction chemistry associated with the solution combustion synthesis. The aim is to synthesize 1 g of the $La_{0.5}A_{0.5}MnO_3$ perovskite; hence, all the calculations are based on 1g of perovskite manufacturing. After completely dissolving the metal nitrates in the deionized water, glycine is added to the mixture as a fuel. The prepared nitrate solution is pre-heated to a predefined temperature (300 to 400˚C) using a temperature-controlled hot plate. At first, due to the heating, the water from the mixture is evaporated, and a viscous gel is formed. After achieving the gel formation, the preheating cause the temperature of the gel to increase up to the auto-ignition point. The auto-ignition and combustion of the gel happen in simultaneous steps, and we can form synthesized $La_{0.5}A_{0.5}MnO_3$ perovskite. The obtained powder is further annealed in a high-temperature muffle furnace up to 1400˚C to attain a crystalline morphology.

## 3. Results and Discussion

Keeping the thermogravimetric analyzer set-up (purchased from SETARAM, France) at the center, a complete reactor set-up is developed in the laboratory to test the combustion synthesized $La_{0.5}A_{0.5}MnO_3$ perovskites (Fig. 2). This setup includes a high-temperature thermogravimetric analyzer (as the heart of the system), gas cylinders for inert Ar and $CO_2$, a vacuum pump, a chiller, mass flow controllers, etc. The platinum crucible is used to hold the powdered $La_{0.5}A_{0.5}MnO_3$ perovskite material. It is a hanging balance, and the mass analysis balance is at the top of the column. Before performing the actual thermochemical reactions, blank experiments are conducted to eliminate the buoyancy effect. Multiple thermal reduction and $CO_2$ splitting steps are conducted at steady and unsteady temperatures. Using the equations reported elsewhere, the mass variations recorded by the TGA are converted into the amount of $O_2$ released, and CO produced.

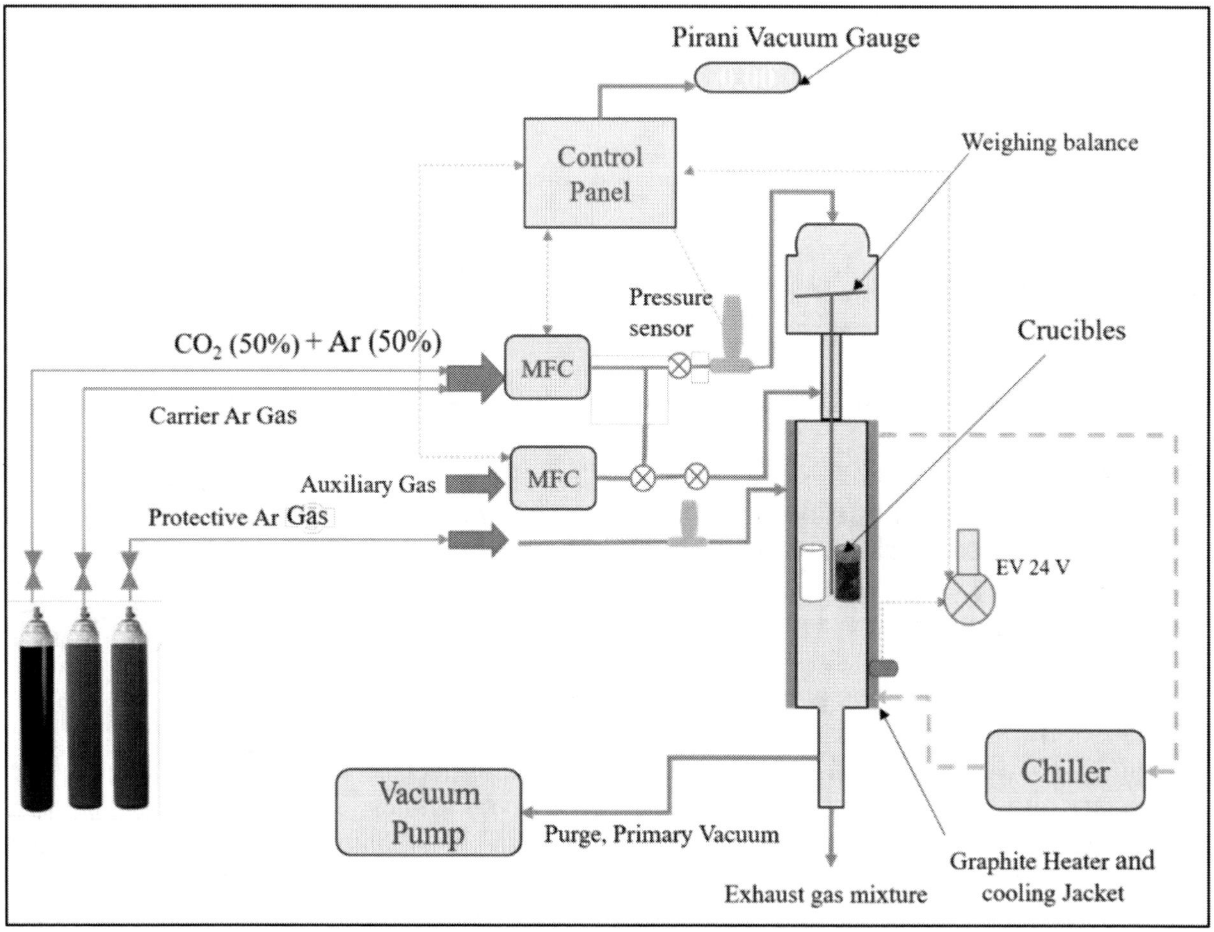


Fig. 2. A thermochemical $CO_2$ splitting reactor set-up.

As explained in the previous section, $La_{0.5}A_{0.5}MnO_3$ perovskites are prepared by the solution combustion method, and the powder obtained is further calcined up to 1400°C in a high-temperature muffle furnace. The calcined powder is further characterized using the powder x-ray diffractometer (PXRD) and scanning electron microscope (SEM). The PXRD peaks indicate the formation of crystalline phase pure $La_{0.5}A_{0.5}MnO_3$ perovskites. The SEM analysis confirms the formation of perovskite porous chunks containing nanoparticles. The preliminary results of the thermochemical experiments indicate that the Sr- and Ba-doped La-Mn-O perovskites have better potential to split $CO_2$ into CO than the Ca- and Mg-doped La-Mn-O perovskites.

## Conclusions

Ca-, Mg-, Sr-, and Ba-doped La-Mn-O perovskites are successfully derived using the glycine-driven solution combustion synthesis. PXRD and SEM analysis confirm the pure phase composition and porous/nanoparticulate morphology of the combustion synthesized $La_{0.5}A_{0.5}MnO_3$ perovskites. All four $La_{0.5}A_{0.5}MnO_3$ perovskites are tested for the thermochemical $CO_2$ splitting reactions using the set-up reported and explained in Fig. 2.

## 5. Acknowledgments

The authors would like to acknowledge the support provided by Qatar University International Research Collaboration Co-funds Grant No. IRCC-2021-002 and the University of Tennessee, Chattanooga, TN, 37421, USA.

## References

[1] Bhosale RR, Mahajani V v. Kinetics of Absorption of Carbon Dioxide in Aqueous Solution of Ethylaminoethanol Modified with N-methyl-2-pyrolidone. Sep Sci Technol 2013;48:2324–37. https://doi.org/10.1080/01496395.2013.805228.
[2] Vaidya PD, Kenig EY. CO2-Alkanolamine Reaction Kinetics: A Review of Recent Studies. Chem Eng Technol 2007;30:1467–74. https://doi.org/10.1002/ceat.200700268.
[3] Villasmil W, Brkic M, Wuillemin D, Meier A, Steinfeld A. Pilot Scale Demonstration of a 100-kWth Solar Thermochemical Plant for the Thermal Dissociation of ZnO. J Sol Energy Eng 2014;136. https://doi.org/10.1115/1.4025512.
[4] Bhosale RR. Thermodynamic efficiency analysis of zinc oxide based solar driven thermochemical H2O splitting cycle: Effect of partial pressure of O2, thermal reduction and H2O splitting temperatures. Int J Hydrogen Energy 2018;43:14915–24. https://doi.org/10.1016/j.ijhydene.2018.06.074.
[5] Bhosale RR. Solar hydrogen production via ZnO/Zn based thermochemical water splitting cycle: Effect of partial reduction of ZnO. Int J Hydrogen Energy 2021;46:4739–48. https://doi.org/10.1016/j.ijhydene.2020.02.135.
[6] Loutzenhiser PG, Meier A, Steinfeld A. Review of the Two-Step H2O/CO2-Splitting Solar Thermochemical Cycle Based on Zn/ZnO Redox Reactions. Materials 2010;3:4922–38. https://doi.org/10.3390/ma3114922.
[7] Loutzenhiser PG, Elena Gálvez M, Hischier I, Graf A, Steinfeld A. CO2 splitting in an aerosol flow reactor via the two-step Zn/ZnO solar thermochemical cycle. Chem Eng Sci 2010;65:1855–64. https://doi.org/10.1016/j.ces.2009.11.025.
[8] Loutzenhiser PG, Steinfeld A. Solar syngas production from CO2 and H2O in a two-step thermochemical cycle via Zn/ZnO redox reactions: Thermodynamic cycle

analysis. Int J Hydrogen Energy 2011;36:12141–7. https://doi.org/10.1016/j.ijhydene.2011.06.128.
[9] Bhosale RR. Thermodynamic efficiency analysis of ZnO/Zn based solar thermochemical CH4 reforming and H2O splitting cycle. Int J Hydrogen Energy 2020;45:5760–71. https://doi.org/10.1016/j.ijhydene.2019.05.191.
[10] Chambon M, Abanades S, Flamant G. Solar thermal reduction of ZnO and SnO2: Characterization of the recombination reaction with O2. Chem Eng Sci 2010;65:3671–80. https://doi.org/10.1016/j.ces.2010.03.005.
[11] Abanades S. CO2 and H2O reduction by solar thermochemical looping using SnO2/SnO redox reactions: Thermogravimetric analysis. Int J Hydrogen Energy 2012;37:8223–31. https://doi.org/10.1016/j.ijhydene.2012.02.158.
[12] ABANADES S, CHARVIN P, LEMONT F, FLAMANT G. Novel two-step SnO2/SnO water-splitting cycle for solar thermochemical production of hydrogen. Int J Hydrogen Energy 2008;33:6021–30. https://doi.org/10.1016/j.ijhydene.2008.05.042.
[13] Bhosale RR, Gupta RB, Shende R v. Thermodynamic evaluation of solar assisted ZnO/Zn thermochemical CO2 splitting cycle. Environ Res 2022;212:113266. https://doi.org/10.1016/j.envres.2022.113266.
[14] Bhosale RR, Shende R v., Gupta RB. <scp> SnO $_2$ </scp> / <scp>SnO</scp> based redox thermochemical <scp> CO $_2$ </scp> splitting cycle: Effect of inert gas flowrate, reduction temperature, and gas separation on the <scp>solar-to-fuel</scp> energy conversion efficiency. Int J Energy Res 2022;46:9267–80. https://doi.org/10.1002/er.7804.
[15] Bhosale RR, Kumar A, Sutar P. Thermodynamic analysis of solar driven SnO2/SnO based thermochemical water splitting cycle. Energy Convers Manag 2017;135:226–35. https://doi.org/10.1016/j.enconman.2016.12.067.
[16] Takalkar G, Bhosale RR, AlMomani F. Sol-gel synthesized NixFe3−xO4 for thermochemical conversion of CO2. Appl Surf Sci 2019;489:693–700. https://doi.org/10.1016/j.apsusc.2019.05.285.
[17] Scheffe JR, McDaniel AH, Allendorf MD, Weimer AW. Kinetics and mechanism of solar-thermochemical H2 production by oxidation of a cobalt ferrite–zirconia composite. Energy Environ Sci 2013;6:963. https://doi.org/10.1039/c3ee23568h.
[18] Kaneko H, Gokon N, Hasegawa N, Tamaura Y. Solar thermochemical process for hydrogen production using ferrites. Energy 2005;30:2171–8. https://doi.org/10.1016/j.energy.2004.08.020.
[19] RV Shende, JA Puszynski, MK Opoku, RR Bhosale. Synthesis of novel ferrite foam material for water-splitting application. Proceedings of NSTI Nanotech conference & expo., 2009, p. 201–4.
[20] Bhosale RR, Kumar A, AlMomani F, Alxneit I. Propylene oxide assisted sol–gel synthesis of zinc ferrite nanoparticles for solar fuel production. Ceram Int 2016;42:2431–8. https://doi.org/10.1016/j.ceramint.2015.10.043.
[21] Diver RB, Miller JE, Allendorf MD, Siegel NP, Hogan RE. Solar Thermochemical Water-Splitting Ferrite-Cycle Heat Engines. J Sol Energy Eng 2008;130. https://doi.org/10.1115/1.2969781.
[22] Bhosale RR, Alxneit I, van den Broeke LLP, Kumar A, Jilani M, Gharbia SS, et al. Sol-Gel Synthesis of Nanocrystalline Ni-Ferrite and Co-Ferrite Redox Materials for

Thermochemical Production of Solar Fuels. MRS Proceedings, 2014, p. 203–8. https://doi.org/10.1557/opl.2014.866.
[23] Agrafiotis C, Roeb M, Sattler C. A review on solar thermal syngas production via redox pair-based water/carbon dioxide splitting thermochemical cycles. Renewable and Sustainable Energy Reviews 2015;42:254–85. https://doi.org/10.1016/j.rser.2014.09.039.
[24] Agrafiotis C, Zygogianni A, Pagkoura C, Kostoglou M, Konstandopoulos AG. Hydrogen production via solar-aided water splitting thermochemical cycles with nickel ferrite: Experiments and modeling. AIChE Journal 2013;59:1213–25. https://doi.org/10.1002/aic.13882.
[25] Scheffe JR, Li J, Weimer AW. A spinel ferrite/hercynite water-splitting redox cycle. Int J Hydrogen Energy 2010;35:3333–40. https://doi.org/10.1016/j.ijhydene.2010.01.140.
[26] Amar VS, Puszynski JA, Shende R v. $H_2$ generation from thermochemical water-splitting using yttria stabilized $NiFe_2O_4$ core-shell nanoparticles. Journal of Renewable and Sustainable Energy 2015;7:023113. https://doi.org/10.1063/1.4915312.
[27] Roeb M, Säck J-P, Rietbrock P, Prahl C, Schreiber H, Neises M, et al. Test operation of a 100kW pilot plant for solar hydrogen production from water on a solar tower. Solar Energy 2011;85:634–44. https://doi.org/10.1016/j.solener.2010.04.014.
[28] Takalkar G, Bhosale RR, AlMomani F, Khraisheh M. Thermocatalytic splitting of CO2 using sol-gel synthesized Co-ferrite redox materials. Fuel 2019;257:115965. https://doi.org/10.1016/j.fuel.2019.115965.
[29] Takalkar G, Bhosale RR. Thermochemical $CO_2$ splitting using a sol-gel–synthesized Mg-ferrite–based redox system. Int J Energy Res 2019:er.4716. https://doi.org/10.1002/er.4716.
[30] Budama VK, Rincon Duarte JP, Roeb M, Sattler C. Potential of solar thermochemical water-splitting cycles: A review. Solar Energy 2023;249:353–66. https://doi.org/10.1016/j.solener.2022.11.001.
[31] Randhir K, Rhodes NR, Li L, AuYeung N, Hahn DW, Mei R, et al. Magnesioferrites for solar thermochemical fuel production. Solar Energy 2018;163:1–15. https://doi.org/10.1016/j.solener.2017.12.006.
[32] Furler P, Scheffe J, Gorbar M, Moes L, Vogt U, Steinfeld A. Solar Thermochemical $CO_2$ Splitting Utilizing a Reticulated Porous Ceria Redox System. Energy & Fuels 2012;26:7051–9. https://doi.org/10.1021/ef3013757.
[33] Bhosale RR, Takalkar G, Sutar P, Kumar A, AlMomani F, Khraisheh M. A decade of ceria based solar thermochemical H2O/CO2 splitting cycle. Int J Hydrogen Energy 2019;44:34–60. https://doi.org/10.1016/j.ijhydene.2018.04.080.
[34] Chueh WC, Falter C, Abbott M, Scipio D, Furler P, Haile SM, et al. High-Flux Solar-Driven Thermochemical Dissociation of $CO_2$ and $H_2O$ Using Nonstoichiometric Ceria. Science (1979) 2010;330:1797–801. https://doi.org/10.1126/science.1197834.
[35] Bhosale RR, Takalkar GD. Nanostructured co-precipitated Ce0.9Ln0.1O2 (Ln = La, Pr, Sm, Nd, Gd, Tb, Dy, or Er) for thermochemical conversion of CO2. Ceram Int 2018;44:16688–97. https://doi.org/10.1016/j.ceramint.2018.06.096.

[36] Takalkar GD, Bhosale RR, Kumar A, AlMomani F, Khraisheh M, Shakoor RA, et al. Transition metal doped ceria for solar thermochemical fuel production. Solar Energy 2018;172:204–11. https://doi.org/10.1016/j.solener.2018.03.022.
[37] le Gal A, Abanades S. Dopant Incorporation in Ceria for Enhanced Water-Splitting Activity during Solar Thermochemical Hydrogen Generation. The Journal of Physical Chemistry C 2012;116:13516–23. https://doi.org/10.1021/jp302146c.
[38] Haeussler A, Abanades S, Julbe A, Jouannaux J, Cartoixa B. Solar thermochemical fuel production from H2O and CO2 splitting via two-step redox cycling of reticulated porous ceria structures integrated in a monolithic cavity-type reactor. Energy 2020;201:117649. https://doi.org/10.1016/j.energy.2020.117649.
[39] Bhosale RR, Kumar A, AlMomani F, Ghosh U, Al-Muhtaseb S, Gupta R, et al. Assessment of Ce Zr Hf O2 based oxides as potential solar thermochemical CO2 splitting materials. Ceram Int 2016;42:9354–62. https://doi.org/10.1016/j.ceramint.2016.02.100.
[40] Hao Y, Yang C-K, Haile SM. Ceria–Zirconia Solid Solutions ($Ce_{1-x}Zr_xO_{2-\delta}$, $x \leq 0.2$) for Solar Thermochemical Water Splitting: A Thermodynamic Study. Chemistry of Materials 2014;26:6073–82. https://doi.org/10.1021/cm503131p.
[41] Sai Gautam G, Stechel EB, Carter EA. A First-Principles-Based Sub-Lattice Formalism for Predicting Off-Stoichiometry in Materials for Solar Thermochemical Applications: The Example of Ceria. Adv Theory Simul 2020;3:2000112. https://doi.org/10.1002/adts.202000112.
[42] Bhosale RR, Kumar A, AlMomani F, Alxneit I. Sol–gel derived CeO2–Fe2O3 nanoparticles: Synthesis, characterization and solar thermochemical application. Ceram Int 2016;42:6728–37. https://doi.org/10.1016/j.ceramint.2016.01.042.
[43] Takalkar G, Bhosale RR, Rashid S, AlMomani F, Shakoor RA, al Ashraf A. Application of Li-, Mg-, Ba-, Sr-, Ca-, and Sn-doped ceria for solar-driven thermochemical conversion of carbon dioxide. J Mater Sci 2020;55:11797–807. https://doi.org/10.1007/s10853-020-04875-1.
[44] Li S, Wheeler VM, Kumar A, Venkataraman MB, Muhich CL, Hao Y, et al. Thermodynamic Guiding Principles for Designing Nonstoichiometric Redox Materials for Solar Thermochemical Fuel Production: Ceria, Perovskites, and Beyond. Energy Technology 2022;10:2000925. https://doi.org/10.1002/ente.202000925.
[45] de la Calle A, Bayon A. Annual performance of a thermochemical solar syngas production plant based on non-stoichiometric CeO2. Int J Hydrogen Energy 2019;44:1409–24. https://doi.org/10.1016/j.ijhydene.2018.11.076.
[46] Takalkar G, Bhosale RR, AlMomani F, Rashid S, Qiblawey H, Saleh Saad MA, et al. Thermochemical splitting of CO2 using solution combustion synthesized lanthanum–strontium–manganese perovskites. Fuel 2021;285:119154. https://doi.org/10.1016/j.fuel.2020.119154.
[47] Scheffe JR, Weibel D, Steinfeld A. Lanthanum–Strontium–Manganese Perovskites as Redox Materials for Solar Thermochemical Splitting of $H_2O$ and $CO_2$. Energy & Fuels 2013;27:4250–7. https://doi.org/10.1021/ef301923h.
[48] Takalkar G, Bhosale RR. Solar thermocatalytic conversion of CO2 using PrxSr(1−x)MnO3−δ perovskites. Fuel 2019;254:115624. https://doi.org/10.1016/j.fuel.2019.115624.

[49] Cooper T, Scheffe JR, Galvez ME, Jacot R, Patzke G, Steinfeld A. Lanthanum Manganite Perovskites with Ca/Sr A-site and Al B-site Doping as Effective Oxygen Exchange Materials for Solar Thermochemical Fuel Production. Energy Technology 2015;3:1130–42. https://doi.org/10.1002/ente.201500226.

[50] Bhosale RR, Kumar A, Ashok A, Sutar P, Takalkar G, Khraisheh M, et al. La-Based Perovskites as Oxygen-Exchange Redox Materials for Solar Syngas Production. MRS Adv 2017;2:3365–70. https://doi.org/10.1557/adv.2017.300.

[51] Haeussler A, Abanades S, Jouannaux J, Julbe A. Non-Stoichiometric Redox Active Perovskite Materials for Solar Thermochemical Fuel Production: A Review. Catalysts 2018;8:611. https://doi.org/10.3390/catal8120611.

[52] Abanades S. Redox Cycles, Active Materials, and Reactors Applied to Water and Carbon Dioxide Splitting for Solar Thermochemical Fuel Production: A Review. Energies (Basel) 2022;15:7061. https://doi.org/10.3390/en15197061.

[53] Takalkar G, Bhosale R, AlMomani F. Combustion synthesized A0.5Sr0.5MnO3-δ perovskites (where, A = La, Nd, Sm, Gd, Tb, Pr, Dy, and Y) as redox materials for thermochemical splitting of CO2. Appl Surf Sci 2019;489:80–91. https://doi.org/10.1016/j.apsusc.2019.05.284.

[54] Bare ZJL, Morelock RJ, Musgrave CB. A Computational Framework to Accelerate the Discovery of Perovskites for Solar Thermochemical Hydrogen Production: Identification of Gd Perovskite Oxide Redox Mediators. Adv Funct Mater 2022;32:2200201. https://doi.org/10.1002/adfm.202200201.

## 153b Emerging Technology in Clean Energy.

*Saurabh Kr Tiwary, Nehil Shreyash, Sukriti Singh*

Carbon Dioxide ($CO_2$) emission has raised major environmental concerns in the last 20 years and scientists are trying to find alternative green energy sources which can replace the pollution-causing fuels being used at present. In addition, efforts channeled towards mitigating the amount of $CO_2$ present in the atmosphere have resulted in a technology which is deemed essential to make the transition to clean energy and it is called Carbon capture and storage (CCS). In combination with CCS, fossil fuel usage has the potential to fuel industrial growth while simultaneously mitigating $CO_2$ emissions. To capture $CO_2$, various chemicals like amine-based solvents are used. Currently, only monoethanolamine (MEA) has been adopted in commercialized process for carbon capture from flue gas stacks in several post-combustion plants. This process has one significant drawback. The MEA must be regularly refilled and re-packed to make up due to losses in the process. Solvent degradation and high volatility are the reasons which cause these losses. Being a corrosive solvent, the recovery of MEA which takes place after capturing the $CO_2$ necessitates a high amount of energy.

These drawbacks of MEA based $CO_2$ capture from the fossil fuel fired fuel gas stack can be rectified by the application of Industrial hydroxides like sodium Hydroxide, Ammonium Hydroxide etc. As per the extent research, the absorption capacity of $CO_2$ in a NaOH solution is greater when compared to that of MEA. One research has shown that the theoretical amount of MEA and NaOH to capture a ton of $CO_2$ is 1.39 and 0.9 tons, respectively. Furthermore, NaOH and other hydroxides are available in replete amounts and at an affordable rate as compared to MEA. However, when compared to MEA, NaOH cannot readily be reconverted or regenerated back to its original form post-capture of $CO_2$. This is primarily due to the byproduct of the reaction between $CO_2$ and NaOH, which is $NaHCO_3$, a compound that shows good solubility in aqueous medium. At a temperature of 333K, it decomposes into $Na_2CO_3$, $H_2O$ and $CO_2$. $Na_2CO_3$ is thermally very stable and starts decomposing (to give $Na_2O$) at temperatures close to 1100K. NaOH's direct source is $Na_2O$, however, this apparent drawback of NaOH serves our purpose well.

Our purpose is to use the NaOH directly for flooding and oil recovery in depleted reservoirs post capture of CO2, removing any need for NaOH recovery. NaOH is often used as an alkali in Alkali-surfactant-polymer (ASP) flooding to produce a conducive environment for micelle formation. With injection of $CO_2$ laden NaOH, not only can the $CO_2$ be sequestered downhole much easily but the presence of $CO_2$ in form of bubbles can help in the production of extra oil due to alteration of the oil miscibility. NaOH can be further substituted with other hydroxides as permitted by the oilfield conditions and parameters to not only enhance $CO_2$ capture but also improve oil recovery from depleted oil fields.

## 107a Statistical Analysis of Production of Bioethanol Produced from *Psidium Gujava* using *S.Cerevisiae*

*Arun Lanka, Vishal Murali*

Fuel is one of the most essential commodities in the world, it has been specifically observed in the recent trends that fossil fuel-based resources have shown a rapid increase in cost owing basically to various reasons. Alternative sources are growing in importance to overcome the different difficulties faced by using fossil fuel as a source. Although solar, and hydrogen fuel cells are growing in importance a simple alternative can be obtained using bio-based/biofuel as a substitute for the current fossil fuels.

Bio-based alternatives and products are seen as a perfect replacement for existing fossil fuels as there are not many technological changes (IC Engine to EV System) associated to adapt to the alternative technology, the biofuel has the advantage of being eco-friendly

The current study focuses on determining the optimum conditions necessary to obtain the maximum yield of bioethanol by fermenting the cellulosic materials obtained from locally available vegetation. For the studies, the cellulosic materials chosen are the leaves of *Psidium guajava* which are readily available in the vicinity and have a relatively decent cellulosic content. The study focuses on utilizing *Saccharomyces Cerevisiae* (commercial baker's yeast) as the microbial species to produce the bioethanol. The study was done to optimize parameters like temperatures, pH levels, and microbial concentration. The initial study was done on a multi-step basis involving various flask scale studies of separate batches of samples to test the yield of ethanol from *Psidium guajava* leaves.

The first stage of the experimentation deals with the pre-treatment of cellulosic materials. In this scenario, the pre-treatment method chosen was washing the leaves with treated water and sun drying them for 15 minutes. The leaves were cut to 0.25cm in size, de-lignified, and immersed in a prepared buffer solution of pH varying from 2 to 8 for 24 hours. The obtained solution is filtered out and sterilized to remove all microbial impurities. Following the sterilization process, the fungal species were inoculated to the fermentation broth and stirred for mixing the content effectively and later placed under static incubation for 24 hours. Samples were then collected and analyzed in a UV-Visible spectrophotometer under a wavelength of 600nm.

Post the initial stage of the process, parameter optimization was carried out with a focus on the temperature and pH parameters. The experimentation was done using the one factor at a time technique and the samples were analyzed for the bioethanol content by weight, of the obtained results the maximum 5 values were chosen for further analysis. The second stage of the process focused on varying the microbial concentrations in the fermentation broth, the main idea was to study the effect of the microbial content on the production of ethanol. The results obtained from the above set of data were analyzed to obtain the optimum set of parameters to produce the bioethanol. The final stage of the work focuses on utilizing the obtained data from the scale studies to analyze its effectivity in the scale-up process which is done using a 3L Laboratory scale Bioreactor.

It was observed that the flask scale analysis for the yield of bioethanol has shown a maximum value at a pH value of 5.8, temperatures of 20°C and 30°C with a microbial concentration of 10% by weight and the yield of bio-ethanol was 16% by weight and 15% by weight respectively and at pH of 6 and temperatures of 40°C and 50°C with a microbial concentration of 10% by weight and the yield of bio-ethanol was 14% and 16% by weight respectively.

All though the yield is quite low in the experimental work, this process shows a promise that normal raw materials and commercial bakers' yeast available over the counter are quite effective in the production of bioethanol. Since a relatively high yield is obtained, one can imply that with further research and development the process can be streamlined to increase the overall production of the bio-ethanol in the near future.

Further studies on the yield based on Design of Experiments and Machine learning algorithms were performed to predict the effect of temperature and pH on the yield of the end product.

SPRING23
+19TH GCPS
A Joint AIChE and CCPS Meeting

# Solar Fuel Production Via Mn-Ferrite Based Water/$CO_2$ Splitting Cycle

**(Principal Author)**
**Rahul Bhosale**
Department of Chemical Engineering,
University of Tennessee, Chattanooga, TN, 37421, USA
rahul-bhosale@utc.edu
rrbhosle1985@gmail.com

**(Co-Authors)**
**Vipul Patil**
Department of Mechanical Engineering,
University of Tennessee, Chattanooga, TN, 37421, USA

**Gabrielle Bennett**
Department of Chemical Engineering,
University of Tennessee, Chattanooga, TN, 37421, USA

Prepared for Presentation at
American Institute of Chemical Engineers
2023 Spring Meeting and 19th Global Congress on Process Safety
Houston, TX, USA
March 12 – 16, 2023

# Solar Fuel Production Via Mn-Ferrite Based Water/$CO_2$ Splitting Cycle

**Rahul R. Bhosale*[,1], Vipul Patil[2], Gabrielle Bennett[1]**

[1]Department of Chemical Engineering, University of Tennessee, Chattanooga, TN, 37421, USA
[2]Department of Mechanical Engineering, University of Tennessee, Chattanooga, TN, 37421, USA

**Keywords:** Mn-ferrite, Thermodynamics, Solar Fuel, Efficiency

**Abstract**

This paper reports the results of a thermodynamic efficiency analysis of an Mn-ferrite-based water/$CO_2$ splitting cycle. A process configuration is developed, and a non-stochiometric thermodynamic model is designed. The efficiency calculations are performed using the thermodynamic data obtained from the HSC Chemistry software. The reduction non-stoichiometry of Mn-ferrite was maintained constant for the entire efficiency analysis.

## 1. Introduction

Metal oxide-based thermochemical splitting of water/$CO_2$ is helpful for the production of solar fuels such as solar $H_2$ and/or solar syngas. Many metal oxides were used in the past for producing solar fuels, such as zinc oxide[1–7], tin oxide[8–13], ferrites[14–29], doped ceria[30–43], and perovskites[44–52]. Most of these metal oxides were utilized to split $H_2O$ to produce $H_2$. However, for syngas (a mixture of $H_2$ and CO) production, it is vital to understand the application of these metal oxides toward $CO_2$ splitting. The conversion of $CO_2$ into fuels has equal importance compared to $CO_2$ capture[53,54]. Mainly the focus of this study is on the use of ferrites for the splitting of $CO_2$. In particular, the application of $MnFe_2O_4$ (Mn-ferrite) for the production of CO from $CO_2$. The reactions applicable in the case of Mn-ferrite-based thermochemical splitting of $CO_2$ are listed below.

$$MnFe_2O_4 \rightarrow MnFe_2O_{4-\delta_{red}} + \frac{\delta_{red}}{2} O_2 \tag{1}$$

$$MnFe_2O_{4-\delta_{red}} + (\delta_{red})CO_2 \rightarrow MnFe_2O_4 + (\delta_{red})CO \tag{2}$$

## 2. Process Flow Configuration & Thermodynamic Model

A process configuration is designed to perform the detailed thermodynamic efficiency analysis of the Mn-ferrite-based $CO_2$ splitting cycle (Fig. 1). This process configuration involves all the essential equipment to be installed in a real-life thermochemical splitting plant. For instance, a reduction chamber, an oxidation chamber, heat exchangers, separators, heaters, a fuel cell, etc.

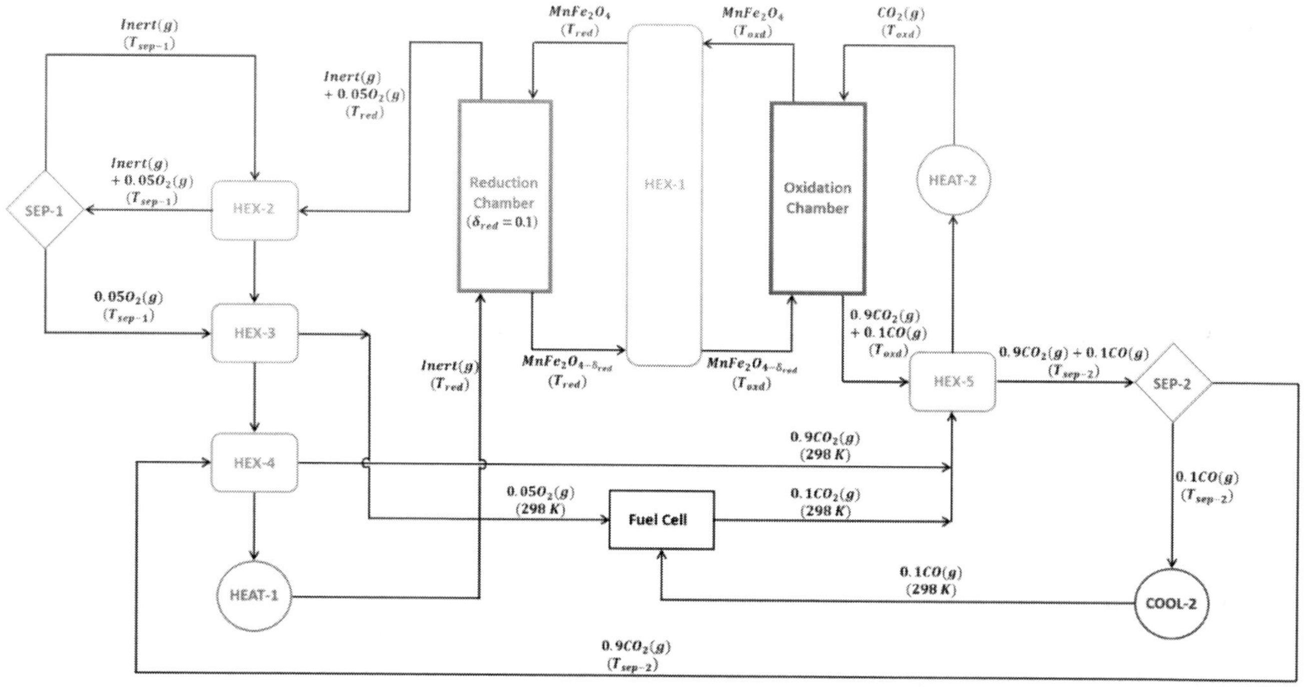


Fig. 1. Schematic of the process model of $MnFe_2O_4$ based CO2 splitting cycle.

**Note: Do not add page numbers. Do not refer to page numbers when referencing different portions of the paper**

A thermodynamic model is developed and used to estimate critical process parameters using this process configuration. Some of the integral model equations are listed below:

$$\eta_{solar-to-fuel} = \frac{\dot{n}_{CO} \times HHV_{CO}}{\dot{Q}_{solar}} \tag{3}$$

$$\dot{Q}_{solar} = \frac{\dot{Q}_{TC}}{\eta_{absoption}} \tag{4}$$

$$\eta_{absorption} = 1 - \left(\frac{\sigma T_{red}^4}{IC}\right) \tag{5}$$

$$\dot{Q}_{TC} = \dot{Q}_{MnF-red} + \dot{Q}_{MnF-sens} + \dot{Q}_{heater-1} + \dot{Q}_{heater-2} + \dot{Q}_{sep-1} + \dot{Q}_{sep-2} + \dot{Q}_{surf} \tag{6}$$

$$\dot{Q}_{MnF-red} = \dot{n}_{MnF} \Delta H|_{MnFe_2O_4 \to MnFe_2O_{4-\delta_{red}} + \frac{\delta_{red}}{2} O_2} \tag{7}$$

$$\dot{Q}_{MnF-sens} = \dot{n}_{MnF} \Delta H|_{MnFe_2O_4 @ T_{oxd} \to MnFe_2O_4 @ T_{red}} \tag{8}$$

## 3. Results and Discussion

At the start of the study, the influence of the inert gas flowrate on the temperature required for the thermal reduction of Mn-ferrite by $\delta_{red}$ = 0.1 is estimated. As shown in Fig. 2, the maximum thermal reduction temperature required to achieve $\delta_{red}$ = 0.1 is recorded for inert gas flowrate equal to 10 times Mn-ferrite's molar flowrate (maintained at 1 mol/s). As the inert gas flowrate over the molar flowrate of the Mn-ferrite ratio enhances from 10 to 100, the temperature required to reduce the Mn-ferrite by $\delta_{red}$ = 0.1 decreases by 140 K.

At the lowest possible thermal reduction temperature equal to 1265K, the critical thermodynamic parameters were estimated by assuming $\delta_{red}$ = 0.1 and by considering 70% heat exchange effectiveness. The thermal energy required to reduce the Mn-ferrite at these operating conditions is estimated at 27.75 kW. The sensible heat energy needed to increase the temperature of Mn-ferrite from oxidation to reduction chamber is calculated to be equal to 51.00 kW. Although heat exchangers are installed to recover and reutilize the heat, additional thermal energy is still needed to close the energy balance. That heat energy is supplied by heater-1 (139.90 kW) and heater-2 (13.59 kW). After estimating the energy penalty associated with the gas separation, the total thermal energy required to drive this cycle is determined as per Eq. (6) and noted to be 271.7 kW. As the re-radiation losses from the process are equal to 13.80 kW, the total solar energy required to drive the cycles is equal to 285.5 kW. Using Eq. (3), the solar-to-fuel energy conversion efficiency is calculated and observed to equal 9.90%.

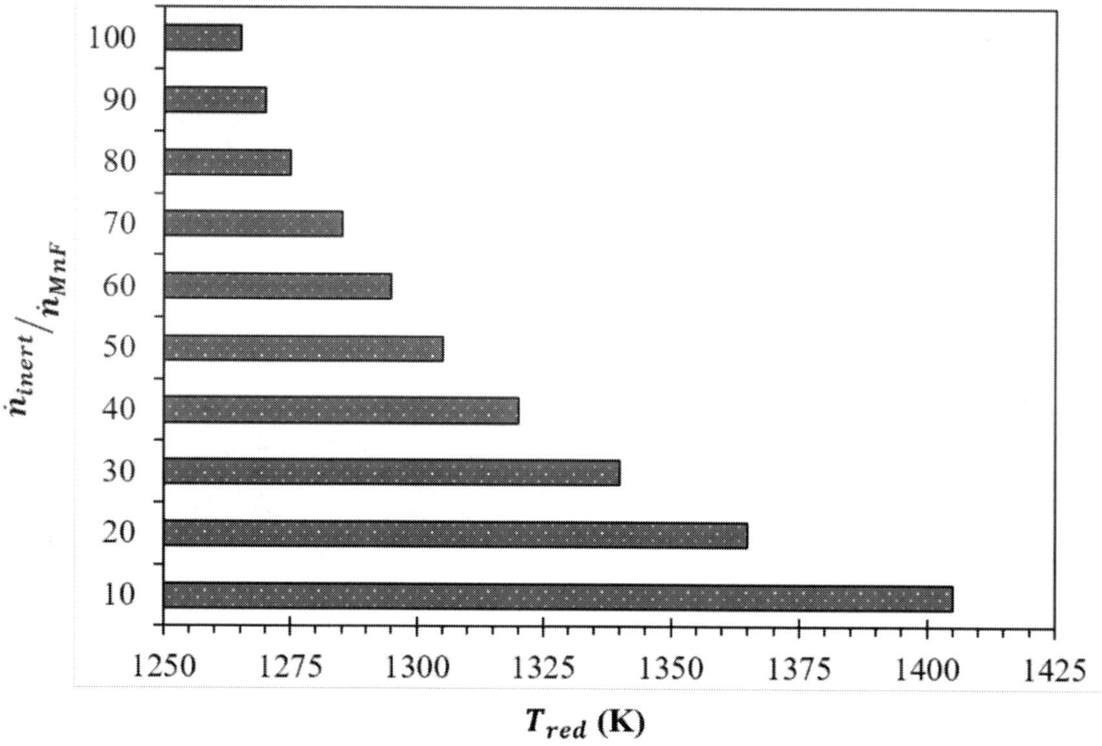


Fig. 2. Effect of $\dot{n}_{inert}/\dot{n}_{MnF}$ on the temperature required for the reduction ($T_{red}$) of $MnFe_2O_4$ ($\delta_{red} = 0.1$).

## 4. Conclusions

A thermodynamic efficiency analysis of the Mn-ferrite-based $CO_2$ splitting cycle is conducted. The results indicate that the rise in the inert gas flow rate significantly influences the thermal reduction temperature. Furthermore, it was also understood that the solar-to-fuel energy conversion efficiency of approximately 10% can be attained at thermal reduction and re-oxidation temperatures of 1265 K and 1000 K, respectively.

## 5. Acknowledgments

The authors would like to acknowledge the support provided by the University of Tennessee, Chattanooga, TN, 37421, USA.

## 6. Nomenclature

| | |
|---|---|
| $C$ | Solar flux concentration ratio, suns |
| $HHV_{CO}$ | Higher heating value of CO, kW |
| HEX-1 | Heat exchanger-1 |
| HEX-2 | Heat exchanger-2 |
| HEX-3 | Heat exchanger-3 |
| HEX-4 | Heat exchanger-4 |
| HEX-5 | Heat exchanger-5 |
| $I$ | Normal beam solar insolation, W/m$^2$ |
| $MO$ | Metal oxide |
| $n$ | Molar amount, mol |

| | |
|---|---|
| $n_{CO,oxd_3}$ | Molar amount of CO at state $oxd_3$, mol |
| $n_{CO,oxd_4}$ | Molar amount of CO at state $oxd_4$, mol |
| $n_{O_2,red_1}$ | Molar amount of $O_2$ at state $red_1$, mol |
| $n_{O_2,red_2}$ | Molar amount of $O_2$ at state $red_2$, mol |
| $\dot{n}$ | Molar flow rate, mol/s |
| $\dot{n}_{inert}$ | Molar flow rate of inert, mol/s |
| $\dot{n}_{MnF}$ | Molar flow rate of $MnFe_2O_4$, mol/s |
| $\dot{n}_{inert}/\dot{n}_{MnF}$ | Ratio of molar flow rates of inert to $MnFe_2O_4$ |
| $\dot{n}_{CO_2}$ | Molar flow rate of $CO_2$, mol/s |
| $\dot{n}_{CO}$ | Molar flow rate of CO, mol/s |
| $\dot{n}_{O_2}$ | Molar flow rate of $O_2$, mol/s |
| $\dot{Q}_{inert-heat}$ | Thermal energy required to heat inert sweep gas, kW |
| $\dot{Q}_{heater-1}$ | Auxiliary thermal energy required to heat inert sweep gas, kW |
| $\dot{Q}_{heater-2}$ | Auxiliary thermal energy required to heat $CO_2$, kW |
| $\dot{Q}_{O_2-cool}$ | Thermal energy released during cooling of $O_2$, kW |
| $\dot{Q}_{solar}$ | Solar energy required to run the cycle, kW |
| $\dot{Q}_{sep-1}$ | Thermal energy required for the operation of separator-1, kW |
| $\dot{Q}_{sep-2}$ | Thermal energy required for the operation of separator-2, kW |
| $\dot{Q}_{surf}$ | Thermal energy losses over the reduction chamber walls, kW |
| $\dot{Q}_{TC}$ | Thermal energy required to run the cycle, kW |
| $\dot{Q}_{MnF-red}$ | Thermal energy required for thermal reduction of $MnFe_2O_4$, kW |
| $\dot{Q}_{MnF-sens}$ | Thermal energy required to heat the $MnFe_2O_4$, kW |
| $\dot{Q}_{MnF-oxd}$ | Thermal energy released during re-oxidation of $MnFe_2O_4$, kW |
| $\dot{Q}_{re-rad}$ | Re-radiation losses from the cycle, kW |
| $R$ | Ideal gas constant (8.314 J/mol·K) |
| $T_0$ | Ambient temperature, K |
| $T_{oxd}$ | Oxidation (splitting) temperature, K |
| $T_{red}$ | Reduction temperature, K |
| $\eta_{absoprtion}$ | Solar energy absorption efficiency, % |
| $\eta_{solar-to-fuel}$ | Solar-to-fuel energy conversion efficiency, % |
| $\delta_{red}$ | Reduction nonstiochiometry |
| $\varepsilon_{gg}$ | Gas-to-gas heat recovery effectiveness |
| $\sigma$ | Stefan – Boltzmann constant, $5.670 \times 10^{-8}$ ($W/m^2 \cdot K^4$) |

## References


[1] Villasmil W, Brkic M, Wuillemin D, Meier A, Steinfeld A. Pilot Scale Demonstration of a 100-kWth Solar Thermochemical Plant for the Thermal Dissociation of ZnO. J Sol Energy Eng 2014;136. https://doi.org/10.1115/1.4025512.

[2] Bhosale RR. Thermodynamic efficiency analysis of zinc oxide based solar driven thermochemical H2O splitting cycle: Effect of partial pressure of O2, thermal reduction and H2O splitting temperatures. Int J Hydrogen Energy 2018;43:14915–24. https://doi.org/10.1016/j.ijhydene.2018.06.074.

[3] Bhosale RR. Solar hydrogen production via ZnO/Zn based thermochemical water splitting cycle: Effect of partial reduction of ZnO. Int J Hydrogen Energy 2021;46:4739–48. https://doi.org/10.1016/j.ijhydene.2020.02.135.

[4] Loutzenhiser PG, Meier A, Steinfeld A. Review of the Two-Step H2O/CO2-Splitting Solar Thermochemical Cycle Based on Zn/ZnO Redox Reactions. Materials 2010;3:4922–38. https://doi.org/10.3390/ma3114922.

[5] Loutzenhiser PG, Elena Gálvez M, Hischier I, Graf A, Steinfeld A. CO2 splitting in an aerosol flow reactor via the two-step Zn/ZnO solar thermochemical cycle. Chem Eng Sci 2010;65:1855–64. https://doi.org/10.1016/j.ces.2009.11.025.

[6] Loutzenhiser PG, Steinfeld A. Solar syngas production from CO2 and H2O in a two-step thermochemical cycle via Zn/ZnO redox reactions: Thermodynamic cycle analysis. Int J Hydrogen Energy 2011;36:12141–7. https://doi.org/10.1016/j.ijhydene.2011.06.128.

[7] Bhosale RR. Thermodynamic efficiency analysis of ZnO/Zn based solar thermochemical CH4 reforming and H2O splitting cycle. Int J Hydrogen Energy 2020;45:5760–71. https://doi.org/10.1016/j.ijhydene.2019.05.191.

[8] Chambon M, Abanades S, Flamant G. Solar thermal reduction of ZnO and SnO2: Characterization of the recombination reaction with O2. Chem Eng Sci 2010;65:3671–80. https://doi.org/10.1016/j.ces.2010.03.005.

[9] Abanades S. CO2 and H2O reduction by solar thermochemical looping using SnO2/SnO redox reactions: Thermogravimetric analysis. Int J Hydrogen Energy 2012;37:8223–31. https://doi.org/10.1016/j.ijhydene.2012.02.158.

[10] ABANADES S, CHARVIN P, LEMONT F, FLAMANT G. Novel two-step SnO2/SnO water-splitting cycle for solar thermochemical production of hydrogen. Int J Hydrogen Energy 2008;33:6021–30. https://doi.org/10.1016/j.ijhydene.2008.05.042.

[11] Bhosale RR, Gupta RB, Shende R v. Thermodynamic evaluation of solar assisted ZnO/Zn thermochemical CO2 splitting cycle. Environ Res 2022;212:113266. https://doi.org/10.1016/j.envres.2022.113266.

[12] Bhosale RR, Shende R v., Gupta RB. <scp> SnO $_2$ </scp> / <scp>SnO</scp> based redox thermochemical <scp> CO $_2$ </scp> splitting cycle: Effect of inert gas flowrate, reduction temperature, and gas separation on the <scp>solar-to-fuel</scp> energy conversion efficiency. Int J Energy Res 2022;46:9267–80. https://doi.org/10.1002/er.7804.

[13] Bhosale RR, Kumar A, Sutar P. Thermodynamic analysis of solar driven SnO2/SnO based thermochemical water splitting cycle. Energy Convers Manag 2017;135:226–35. https://doi.org/10.1016/j.enconman.2016.12.067.

[14] Takalkar G, Bhosale RR, AlMomani F. Sol-gel synthesized NixFe3−xO4 for thermochemical conversion of CO2. Appl Surf Sci 2019;489:693–700. https://doi.org/10.1016/j.apsusc.2019.05.285.

[15] Scheffe JR, McDaniel AH, Allendorf MD, Weimer AW. Kinetics and mechanism of solar-thermochemical H2 production by oxidation of a cobalt ferrite–zirconia composite. Energy Environ Sci 2013;6:963. https://doi.org/10.1039/c3ee23568h.

[16] Kaneko H, Gokon N, Hasegawa N, Tamaura Y. Solar thermochemical process for hydrogen production using ferrites. Energy 2005;30:2171–8. https://doi.org/10.1016/j.energy.2004.08.020.

[17] RV Shende, JA Puszynski, MK Opoku, RR Bhosale. Synthesis of novel ferrite foam material for water-splitting application. Proceedings of NSTI Nanotech conference & expo., 2009, p. 201–4.
[18] Bhosale RR, Kumar A, AlMomani F, Alxneit I. Propylene oxide assisted sol–gel synthesis of zinc ferrite nanoparticles for solar fuel production. Ceram Int 2016;42:2431–8. https://doi.org/10.1016/j.ceramint.2015.10.043.
[19] Diver RB, Miller JE, Allendorf MD, Siegel NP, Hogan RE. Solar Thermochemical Water-Splitting Ferrite-Cycle Heat Engines. J Sol Energy Eng 2008;130. https://doi.org/10.1115/1.2969781.
[20] Bhosale RR, Alxneit I, van den Broeke LLP, Kumar A, Jilani M, Gharbia SS, et al. Sol-Gel Synthesis of Nanocrystalline Ni-Ferrite and Co-Ferrite Redox Materials for Thermochemical Production of Solar Fuels. MRS Proceedings, 2014, p. 203–8. https://doi.org/10.1557/opl.2014.866.
[21] Agrafiotis C, Roeb M, Sattler C. A review on solar thermal syngas production via redox pair-based water/carbon dioxide splitting thermochemical cycles. Renewable and Sustainable Energy Reviews 2015;42:254–85. https://doi.org/10.1016/j.rser.2014.09.039.
[22] Agrafiotis C, Zygogianni A, Pagkoura C, Kostoglou M, Konstandopoulos AG. Hydrogen production via solar-aided water splitting thermochemical cycles with nickel ferrite: Experiments and modeling. AIChE Journal 2013;59:1213–25. https://doi.org/10.1002/aic.13882.
[23] Scheffe JR, Li J, Weimer AW. A spinel ferrite/hercynite water-splitting redox cycle. Int J Hydrogen Energy 2010;35:3333–40. https://doi.org/10.1016/j.ijhydene.2010.01.140.
[24] Amar VS, Puszynski JA, Shende R v. $H_2$ generation from thermochemical water-splitting using yttria stabilized $NiFe_2O_4$ core-shell nanoparticles. Journal of Renewable and Sustainable Energy 2015;7:023113. https://doi.org/10.1063/1.4915312.
[25] Roeb M, Säck J-P, Rietbrock P, Prahl C, Schreiber H, Neises M, et al. Test operation of a 100kW pilot plant for solar hydrogen production from water on a solar tower. Solar Energy 2011;85:634–44. https://doi.org/10.1016/j.solener.2010.04.014.
[26] Takalkar G, Bhosale RR, AlMomani F, Khraisheh M. Thermocatalytic splitting of CO2 using sol-gel synthesized Co-ferrite redox materials. Fuel 2019;257:115965. https://doi.org/10.1016/j.fuel.2019.115965.
[27] Takalkar G, Bhosale RR. Thermochemical $CO_2$ splitting using a sol-gel–synthesized Mg-ferrite–based redox system. Int J Energy Res 2019:er.4716. https://doi.org/10.1002/er.4716.
[28] Budama VK, Rincon Duarte JP, Roeb M, Sattler C. Potential of solar thermochemical water-splitting cycles: A review. Solar Energy 2023;249:353–66. https://doi.org/10.1016/j.solener.2022.11.001.
[29] Randhir K, Rhodes NR, Li L, AuYeung N, Hahn DW, Mei R, et al. Magnesioferrites for solar thermochemical fuel production. Solar Energy 2018;163:1–15. https://doi.org/10.1016/j.solener.2017.12.006.

[30] Furler P, Scheffe J, Gorbar M, Moes L, Vogt U, Steinfeld A. Solar Thermochemical $CO_2$ Splitting Utilizing a Reticulated Porous Ceria Redox System. Energy & Fuels 2012;26:7051–9. https://doi.org/10.1021/ef3013757.
[31] Bhosale RR, Takalkar G, Sutar P, Kumar A, AlMomani F, Khraisheh M. A decade of ceria based solar thermochemical H2O/CO2 splitting cycle. Int J Hydrogen Energy 2019;44:34–60. https://doi.org/10.1016/j.ijhydene.2018.04.080.
[32] Chueh WC, Falter C, Abbott M, Scipio D, Furler P, Haile SM, et al. High-Flux Solar-Driven Thermochemical Dissociation of $CO_2$ and $H_2O$ Using Nonstoichiometric Ceria. Science (1979) 2010;330:1797–801. https://doi.org/10.1126/science.1197834.
[33] Bhosale RR, Takalkar GD. Nanostructured co-precipitated Ce0.9Ln0.1O2 (Ln = La, Pr, Sm, Nd, Gd, Tb, Dy, or Er) for thermochemical conversion of CO2. Ceram Int 2018;44:16688–97. https://doi.org/10.1016/j.ceramint.2018.06.096.
[34] Takalkar GD, Bhosale RR, Kumar A, AlMomani F, Khraisheh M, Shakoor RA, et al. Transition metal doped ceria for solar thermochemical fuel production. Solar Energy 2018;172:204–11. https://doi.org/10.1016/j.solener.2018.03.022.
[35] le Gal A, Abanades S. Dopant Incorporation in Ceria for Enhanced Water-Splitting Activity during Solar Thermochemical Hydrogen Generation. The Journal of Physical Chemistry C 2012;116:13516–23. https://doi.org/10.1021/jp302146c.
[36] Haeussler A, Abanades S, Julbe A, Jouannaux J, Cartoixa B. Solar thermochemical fuel production from H2O and CO2 splitting via two-step redox cycling of reticulated porous ceria structures integrated in a monolithic cavity-type reactor. Energy 2020;201:117649. https://doi.org/10.1016/j.energy.2020.117649.
[37] Bhosale RR, Kumar A, AlMomani F, Ghosh U, Al-Muhtaseb S, Gupta R, et al. Assessment of Ce Zr Hf O2 based oxides as potential solar thermochemical CO2 splitting materials. Ceram Int 2016;42:9354–62. https://doi.org/10.1016/j.ceramint.2016.02.100.
[38] Hao Y, Yang C-K, Haile SM. Ceria–Zirconia Solid Solutions ($Ce_{1-x}Zr_xO_{2-\delta}$, $x \leq 0.2$) for Solar Thermochemical Water Splitting: A Thermodynamic Study. Chemistry of Materials 2014;26:6073–82. https://doi.org/10.1021/cm503131p.
[39] Sai Gautam G, Stechel EB, Carter EA. A First-Principles-Based Sub-Lattice Formalism for Predicting Off-Stoichiometry in Materials for Solar Thermochemical Applications: The Example of Ceria. Adv Theory Simul 2020;3:2000112. https://doi.org/10.1002/adts.202000112.
[40] Bhosale RR, Kumar A, AlMomani F, Alxneit I. Sol–gel derived CeO2–Fe2O3 nanoparticles: Synthesis, characterization and solar thermochemical application. Ceram Int 2016;42:6728–37. https://doi.org/10.1016/j.ceramint.2016.01.042.
[41] Takalkar G, Bhosale RR, Rashid S, AlMomani F, Shakoor RA, al Ashraf A. Application of Li-, Mg-, Ba-, Sr-, Ca-, and Sn-doped ceria for solar-driven thermochemical conversion of carbon dioxide. J Mater Sci 2020;55:11797–807. https://doi.org/10.1007/s10853-020-04875-1.
[42] Li S, Wheeler VM, Kumar A, Venkataraman MB, Muhich CL, Hao Y, et al. Thermodynamic Guiding Principles for Designing Nonstoichiometric Redox Materials for Solar Thermochemical Fuel Production: Ceria, Perovskites, and Beyond. Energy Technology 2022;10:2000925. https://doi.org/10.1002/ente.202000925.

[43] de la Calle A, Bayon A. Annual performance of a thermochemical solar syngas production plant based on non-stoichiometric CeO2. Int J Hydrogen Energy 2019;44:1409–24. https://doi.org/10.1016/j.ijhydene.2018.11.076.

[44] Takalkar G, Bhosale RR, AlMomani F, Rashid S, Qiblawey H, Saleh Saad MA, et al. Thermochemical splitting of CO2 using solution combustion synthesized lanthanum–strontium–manganese perovskites. Fuel 2021;285:119154. https://doi.org/10.1016/j.fuel.2020.119154.

[45] Scheffe JR, Weibel D, Steinfeld A. Lanthanum–Strontium–Manganese Perovskites as Redox Materials for Solar Thermochemical Splitting of $H_2O$ and $CO_2$. Energy & Fuels 2013;27:4250–7. https://doi.org/10.1021/ef301923h.

[46] Takalkar G, Bhosale RR. Solar thermocatalytic conversion of CO2 using PrxSr(1−x)MnO3−δ perovskites. Fuel 2019;254:115624. https://doi.org/10.1016/j.fuel.2019.115624.

[47] Cooper T, Scheffe JR, Galvez ME, Jacot R, Patzke G, Steinfeld A. Lanthanum Manganite Perovskites with Ca/Sr A-site and Al B-site Doping as Effective Oxygen Exchange Materials for Solar Thermochemical Fuel Production. Energy Technology 2015;3:1130–42. https://doi.org/10.1002/ente.201500226.

[48] Bhosale RR, Kumar A, Ashok A, Sutar P, Takalkar G, Khraisheh M, et al. La-Based Perovskites as Oxygen-Exchange Redox Materials for Solar Syngas Production. MRS Adv 2017;2:3365–70. https://doi.org/10.1557/adv.2017.300.

[49] Haeussler A, Abanades S, Jouannaux J, Julbe A. Non-Stoichiometric Redox Active Perovskite Materials for Solar Thermochemical Fuel Production: A Review. Catalysts 2018;8:611. https://doi.org/10.3390/catal8120611.

[50] Abanades S. Redox Cycles, Active Materials, and Reactors Applied to Water and Carbon Dioxide Splitting for Solar Thermochemical Fuel Production: A Review. Energies (Basel) 2022;15:7061. https://doi.org/10.3390/en15197061.

[51] Takalkar G, Bhosale R, AlMomani F. Combustion synthesized A0.5Sr0.5MnO3-δ perovskites (where, A = La, Nd, Sm, Gd, Tb, Pr, Dy, and Y) as redox materials for thermochemical splitting of CO2. Appl Surf Sci 2019;489:80–91. https://doi.org/10.1016/j.apsusc.2019.05.284.

[52] Bare ZJL, Morelock RJ, Musgrave CB. A Computational Framework to Accelerate the Discovery of Perovskites for Solar Thermochemical Hydrogen Production: Identification of Gd Perovskite Oxide Redox Mediators. Adv Funct Mater 2022;32:2200201. https://doi.org/10.1002/adfm.202200201.

[53] Bhosale RR, Mahajani V v. Kinetics of Absorption of Carbon Dioxide in Aqueous Solution of Ethylaminoethanol Modified with N-methyl-2-pyrolidone. Sep Sci Technol 2013;48:2324–37. https://doi.org/10.1080/01496395.2013.805228.

[54] Vaidya PD, Kenig EY. CO2-Alkanolamine Reaction Kinetics: A Review of Recent Studies. Chem Eng Technol 2007;30:1467–74. https://doi.org/10.1002/ceat.200700268.

# 107c Managing the Risky Business of Scaling Sustainable Technology

*Michael Schultz*

Developing a new technology is inherently filled with risk, and a key objective of any successful scaleup effort is mitigating and reducing risk while managing timeline and budget. There are three key components to this approach:

- Assess Risk. A risk assessment should be completed in the early stages of scaleup, evaluating risk areas such as technical, commercial, supply chain, etc. Scoring each risk by impact and probability provides an understanding of the most critical risks, so that these can be prioritized with mitigation steps outlined to support the scaleup effort. This risk assessment should be re-evaluated periodically to assess the impact of mitigation steps of addressing high risk areas, and to determine if any changes internal or externally have changes the risk factors.
- Decouple scaleup parameters. Next, it is important to decouple and prioritize the scaleup parameters. This effort is informed by the risk assessment. The lab scale is best for evaluating the intrinsic parameters associated with the chemistry/biology of the system, while engineering parameters are better evaluated using calculations and modeling at early stages, with data generation at larger scales. We can use this approach to develop a scaleup plan in which we define critical objectives for each stage.
- Develop multi-scale data in parallel. Finally, multi-scale data is critical to any scaleup effort. A very sequential approach to scaleup—aiming to hit commercial targets at each scale before moving to the next--can reduce risk significantly, but at the expense of timeline. Timeline is a key driver for any new technology, so an alternate approach is to make strategic risk decisions and do much of this work in parallel—advancing to the next scale even while working on proving out commercial targets at the smaller scale. With the risk assessment in place, and a clear scaleup plan, this risk can be managed while accelerating the timeline.

This talk will elaborate on this approach, illustrated by real world examples in which these principles have been applied to effectively manage risk during the scaleup of sustainable fuel and chemical technologies.

# AUTHOR INDEX

**AIChE**
120 Wall Street, FL 23
New York, NY 10005-4020

ISBN 978-1-7138-7934-3

9 781713 879343

CURRAN ASSOCIATES INC.
proceedings
.com